Confusion often arises when comparing the concepts of primitive ape-man, modern man created in the image of God, and the biblical creative days. At first glance, the perspectives of science, theology, and geology seem so vastly different that they appear irreconcilable. With each new scientific or archaeological discovery, the gap between these viewpoints can seem to widen even further. Yet, this paper seeks to bridge that divide by exploring the possibility that the differences are not as great as they appear. By stepping outside the rigid "boxes" that each discipline has built around itself, we may find that science, theology, and geology share more common ground than one might expect.

Carl Sagan, in his book *Cosmos*, candidly acknowledged:

"The fossil evidence could be consistent with the idea of a Great Designer." — *(New York, 1980), p. 29.*

For the sake of discussion, let us take the premise that God—the Grand Designer—does indeed exist. With this assumption as our foundation, we can now move forward in our exploration.

Biblically speaking, there are seven things we know about God:

1. **God's thoughts and ways are far higher than men.**
   Isaiah 55:8–9 (KJV)

"For my thoughts are not your thoughts, neither are your ways my ways, saith the LORD. For as the heavens are higher than the earth, so are my ways higher than your ways, and my thoughts than your thoughts."

2. **God's intellect and abilities surpass human understanding.**
   1 Samuel 16:7 (ASV)

"But Jehovah said unto Samuel, look not on his countenance, or on the height of his stature; because I have rejected him: for Jehovah seeth not as man seeth; for man looketh on the outward appearance…"

3. **God's knowledge exceeds all human and earthly limits.**
   Daniel 4:35 (Byington)

"All who live on earth count for nothing, and he does as he pleases with the legions of the sky and with those who live on earth; and there is no one who shall check his hand and say to him 'What are you doing?'"

4. **God is all-powerful. His energy and authority extend everywhere, even to the farthest reaches of the universe.**
   Isaiah 40:26 (Reference Bible) "Raise your eyes high up and see. Who has created these things? It is the One who is bringing forth the army of them even by number, all of whom he calls even by name. Due to the abundance of dynamic energy, he also being vigorous in power, not one of them is missing."

5. **God cannot lie.**
   Titus 1:2 (KJV) "In hope of eternal life, which God, that cannot lie, promised before the world began."

6. **God's purpose will prevail, and no one can derail it.**
   Isaiah 46:10 (KJV) "Declaring the end from the beginning, and from ancient times the things that are not yet done, saying, My counsel shall stand, and I will do all my pleasure."

7. **God lives forever; He cannot die.**
   Revelation 4:9–11 (KJV) "And when those beasts give glory and honour and thanks to him that sat on the throne, who liveth for ever and ever…"

   > **Greek Interlinear - Revelation 4:9-11**
   > And whenever will give the living [creatures] glory and honor and thanksgiving to the one upon the throne **to the one living into the ages of the ages.**

Illustration: God's Different Values for Time

When we examine Scripture carefully, it becomes clear that God does not assign the same value to time that humans do. What may seem like a very long period to mankind can be nothing more than a moment to the Creator, while what seems like a moment to us may hold eternal weight in His plan.

1. **A thousand years can be like a single day.**
   2 Peter 3:8 (KJV) reminds us:

   "But, beloved, be not ignorant of this one thing, that one day is with the Lord as a thousand years, and a thousand years as one day."
   To God, vast stretches of human history can pass as effortlessly as a single day passes for us. This suggests that geological or archaeological timelines, while staggering to us, are easily within His scope of governance.

2. **A single day can carry eternal significance.**
   Consider the day of Jesus' death. To human eyes, it was only one day in history. Yet from God's perspective, that one day carried the weight of redemption for all mankind across all ages and into the future. Here, the span of time (just hours) was less important than the eternal purpose accomplished.

3. **God uses symbolic time periods to teach.**
   In Genesis, the "days" of creation are presented as sequential periods of creative activity. Whether these "days" are taken literally as 24-hour periods or symbolically as long epochs of creative work, the central truth remains: God measured time by His creative acts, not by human clocks. Similarly, prophetic passages often use "days" to represent years (e.g., Numbers 14:34; Ezekiel 4:6), showing that God can assign time values according to His purpose.

4. **Human lifespans are brief by comparison.**
   Psalm 90:4, 10 (KJV) says: "For a thousand years in thy sight are but as yesterday when it is past, and as a watch in the night… The days of our years are threescore years and ten; and if by reason of strength they be fourscore years…"

   What seems like a full, long life to us is but a fleeting moment to God. Yet He still values those moments deeply because they unfold within His eternal design.

Consider what was revealed regarding the "day" of Adam's death.

   American Standard Version - Genesis 2:17 " but of the tree of the knowledge of good and evil, thou shalt not eat of it: *for in the day* that thou eatest thereof thou shalt surely die."

Adam and Eve did not die during a literal 24-hour day. Adam lived to be much older.

   King James Version - Genesis 5:5 "And all the days that Adam lived were *nine hundred and thirty years*: and he died."

Why didn't Adam die within a 24-hour period of time? The answer is found at 2 Peter 3:8.

   King James Version - 2 Peter 3:8 "But, beloved, be not ignorant of this one thing, that one day *is* with the Lord as a thousand years, and *a thousand years as one day*."

   Adam did die within the definition of this specific time period or "day" according to 2 Peter 3:8.

God does not measure time the way humans do. He assigns value to time not by its length but by its purpose. If this is true, then the length of the biblical creative days doesn't need to be rigidly defined in terms of human chronology. They could be vast epochs of creative work — each serving its own divine purpose. To insist that they *must* fit into one human definition is to impose our mortal urgency upon an eternal God. Man, whose existence is brief and whose days are numbered, naturally obsesses over time. But for the Creator who cannot die, time is not a limitation. It is simply a medium through which His purpose unfolds.

This perspective allows us to step outside the narrow boxes of theology, geology, and science, and see that the differences between them may not be as irreconcilable as they seem to appear. Geology's billions of years, science's fossil record, and theology's creative days need not cancel each other out. Instead, they may represent different vantage points on the same truth: that the universe was designed, ordered, and guided by a timeless Intelligence.

Carl Sagan himself admitted that "the fossil evidence could be consistent with the idea of a Great Designer." If even a scientist of his stature could acknowledge that possibility, perhaps we too should resist drawing hard lines where even Scripture itself allows for breadth and flexibility.

Thinking outside of the box: Creative Days

What about creative days mentioned in the book of Genesis chapters 1 and 2?

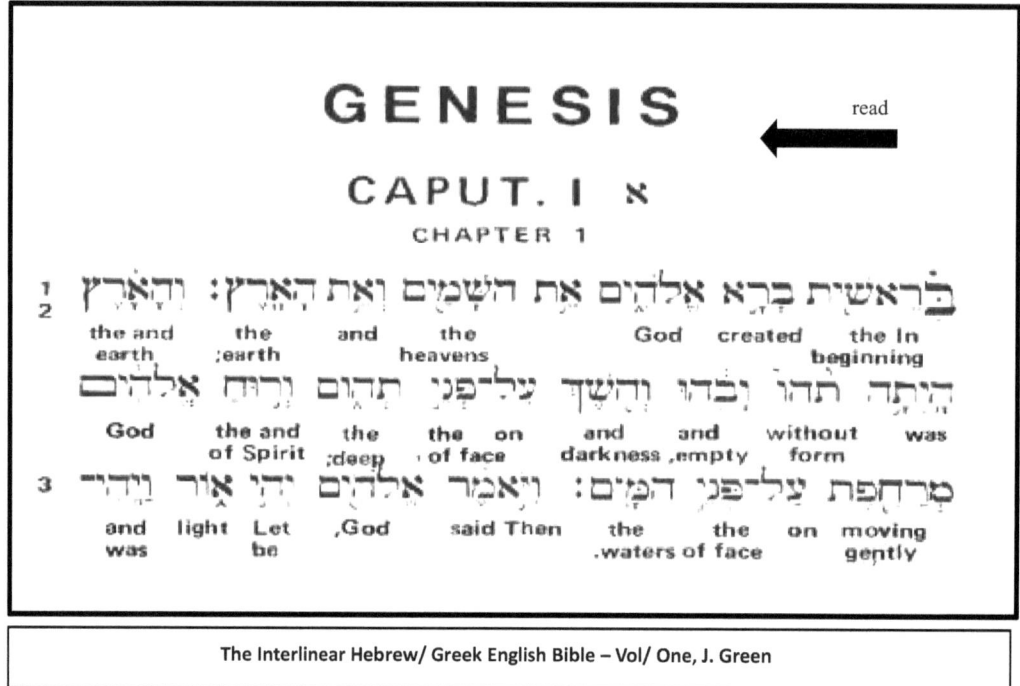

The Interlinear Hebrew/ Greek English Bible – Vol/ One, J. Green

King James Version - Genesis 1:1 In the beginning God created the heaven and the earth.

Geologists date Earth's rocks and fossils through radiometric dating, stratigraphy, and other sciences that give us timeframes of millions and billions of years. Yet astronomy, particularly with advanced space telescopes like Hubble and James Webb, keeps pushing back our understanding of the Universe's age and complexity. Some galaxies appear to have formed earlier than expected, challenging prior models.

It raises the thought: what if the material "stuff" of the Universe—the raw elements—predates what geology assumes about Earth's beginning? In other words, the atoms that make up Earth may have existed long before they were ever drawn together into this planet. That makes the "age of Earth" a relative measure—measured by when it became Earth, not when its materials came into existence.

So, while Earth's geologic timeline might be fixed by formation events, the Universe itself may be much older, and the building blocks of Earth older still. This leaves open the possibility that what we consider "the beginning" is more layered than we think—matter, stars, planets, life, all beginning at different stages, according to different clocks.

Rethinking the Age of Earth and the Universe

For more than a century, geologists and astronomers have worked to measure the age of Earth and the Universe. The prevailing scientific consensus places Earth at roughly 4.5 billion years old, a figure derived from radiometric dating of rocks and meteorites. Likewise, the Universe is generally said to be about 13.8 billion years old, a conclusion drawn from cosmic background radiation and models of expansion. These numbers, while carefully studied, are still subject to debate. As technology improves, especially with the advancement of powerful space telescopes, new evidence has emerged those challenges earlier assumptions about time, origins, and the sequence of creation.

One striking tension lies in the difference between geology and astronomy. Geologists measure Earth's "age" based on when it solidified as a planet, yet this leaves out the fact that its raw materials—the carbon, iron, silicon, and oxygen that make up its surface and core—are far older. Stars lived and died long before Earth ever formed, scattering these elements across space. When telescopes like Hubble and James Webb allow us to peer billions of light-years into the past, we see galaxies forming earlier than expected, pushing the Universe's timeline further back than once thought. If the Universe itself is older than scientists assumed, then it follows that the building blocks of Earth, too, stretch farther into antiquity than the neat boundaries of the geologic eras suggest.

This possibility invites us to reconsider how we define "age." Is Earth 4.5 billion years old simply because that is when it took shape as a planet? Or is its essence—the matter from which it is made—much older, reaching back into the earliest furnaces of the Universe? In this light, the familiar geologic timescale might not be the final word but rather a narrow slice of a much grander chronology. The debate is not about discrediting science, but about realizing that human measurements, however precise, are still provisional, always subject to revision as our tools and perspectives grow.

Ultimately, the study of origins—whether of Earth or of the cosmos—reminds us of humility. What we think we know today may be overturned tomorrow, not because truth has changed, but because our vision has widened. If the Universe is indeed older and deeper than once imagined, then the Earth and everything on it are part of a story far grander than our current timelines can contain.

Adam and the Question of Time

When we trace the Biblical genealogy literally, beginning with Adam, the timeline places humanity's origin around 4026 B.C.E. For many Bible readers, this is taken as the start of human history. Yet when compared with the scientific record—fossils, radiometric dating, and archaeological finds—the age of humanity is pushed back tens of thousands, even hundreds of thousands of years. On the surface, these two frameworks appear irreconcilable. How can one reconcile a 6,000-year Biblical timeline with a scientific timeline that spans millions?

The problem often comes down to interpretation. Science and theology are not always asking the same questions. Science attempts to measure time through physical processes—decay of elements, movement of tectonic plates, or the expansion of the universe and even through evolution processes. Theology, on the other hand, is not primarily concerned with dating rocks or galaxies; it is concerned with the meaning of humanity's origin, our relationship with God, and the moral framework in which life unfolds. From this view, the "time" of Adam is not necessarily the same as the "time" of geology. Adam's placement at 4026 B.C.E. marks the beginning of humanity's history with God, not necessarily the appearance of the first Hominid/ Hominin's as science would define it.

Some theologians argue that Adam may represent the first man chosen by God to bear His image in a spiritual sense, rather than the very first biological erect life form. Others hold firmly to a literal view, believing that the genealogies provide a precise historical timeline. Both perspectives highlight the tension: one seeks harmony with scientific evidence, while the other emphasizes fidelity to the Biblical record. What remains undeniable is that the Biblical account elevates Adam beyond biology—he is not just a man of clay and bone, but the figure through whom sin, responsibility, and hope for redemption entered human history.

The gap, then, may not be so much about time as about whom we are talking about. Science tells us how old things are; Scripture tells us why they matter. While the genealogical line from Adam to today suggests only a few thousand years, the essence of the Biblical message is not to measure the Earth's strata but to reveal God's plan for mankind. Reconciling the two may be difficult—even impossible if both are read rigidly—but recognizing their distinct purposes and our definitions of who is related to whom, allows them to stand side by side without needing to cancel each other out.

> When the impossible has been eliminated, all that remains no matter how improbable is possible. *Arthur Conan Doyle*

The Bible says that man was created in God's image (day 6). Why is that important?

The Interlinear Hebrew/ Greek English Bible – Vol/ One J. Green

Early Man or Merely Apes?

The scientific community often speaks of "early man," suggesting that humanity gradually emerged from a line of ape-like ancestors through a long process of evolution. Yet from a Biblical standpoint, this idea is deeply problematic. Genesis makes a clear distinction: man was created in God's image (Genesis 1:26, 27). If we take that seriously, then whatever came before Adam was not truly man at all but another form of animal life. The so-called "early man" was in fact an ape, created as part of God's diverse living creation, but never intended to be the bearer of His image. This is why, despite over a century of research and fossil discoveries, the much sought-after "missing link" remains elusive—because no such link exists.

Fossil records reveal that around four million years ago certain primates developed a more upright posture. At first, they alternated between walking on two legs and moving on all fours, and over time their anatomy allowed them to walk more regularly on their hind legs. This bipedalism freed their hands for grasping, carrying, and manipulating their environment. Yet none of this made them human in the Biblical sense. Intelligence, creativity, and social behavior may have grown within these primates, but what defines humanity—the spiritual capacity to know God, to bear His likeness, and to live within His moral framework—was entirely absent.

The Bible itself does not comment on the extinction of these ape-like hominoids, nor does it need to. Scripture's purpose is not to trace the biological fate of primates but to highlight God's plan for mankind. Revelation 4:11 reminds us that "all things were created because of [God's] will." If that is the case, then the existence of these creatures had a role within His creation, even if they were never intended to be human. They may have filled an ecological niche, or perhaps their existence served as part of God's progressive creative activity leading up to the formation of Adam, the first true man in His image.

Seen in this light, the confusion between ape and man dissolves. What science calls "early man" was never man at all. Humanity began not when primates learned to walk upright, but when God breathed life into Adam and established him as His image-bearer. This distinction not only aligns with the Biblical account but also explains why the supposed evolutionary bridge between apes and humans has remained an unproven theory rather than a demonstrated fact.

King James Version - Revelation 4:11 "Thou art worthy, O Lord, to receive glory and honour and power: for thou hast created all things, and for thy pleasure they are and were created."

When that purpose, pleasure, or activity had been fulfilled, God allowed the ape-like hominoids to pass into extinction. They had served their role in the grand design of creation, but they were never the true reflection of the Creator, nor the vessel through which His spirit would flow. They were living creatures, remarkable in form and function, yet bound to the limits of the natural world.

Then, after a span of time had run its course, God turned to His next and greater creative work. Out of the dust of the ground, He formed man ['a·dham'], shaping not just a living being but a bearer of His likeness. Unlike the animals, man was endowed with moral awareness, creative capacity, and the ability to enter into fellowship with his Maker. In man, God's image shone forth—reflecting reason, conscience, love, and the potential for everlasting communion with the divine.

Thus, humanity was not an accident of evolution nor the mere refinement of animal stock, but a new creation, deliberately set apart. The extinction of those earlier hominoids was not a loss but a transition, clearing the stage for the arrival of the one creature uniquely designed to walk with God, to rule over the earth in His name, and to share in His eternal purpose.

> There is nothing more deceptive than an obvious fact. **Arthur Conan Doyle**

The most evident truth is that man ('a·dham′) was not created as a physical replica of his Grand Creator. The Scriptures are plain: *"No man hath seen God at any time"* (1 John 4:12, KJV). If no human has ever seen God, then the Hebrew word translated "image" must mean something deeper than mere outward appearance. It cannot be a matter of physical likeness, for as Jesus himself declared, *"God is a Spirit"* (John 4:24). Unlike man, who is flesh and blood, God possesses a spiritual body, one that no earthly eyes have ever beheld. Therefore, any comparison between man's physical form and God's essence is impossible.

The Bible itself points us toward a richer meaning. To be made in God's image is to be endowed with certain moral, intellectual, and spiritual qualities that reflect the Creator's own nature. For instance, Scripture tells us that *"God is love"* (1 John 4:8). In a similar way, man was created with the capacity to love selflessly, to seek justice (Deut. 32:4), and to exercise a conscience that discerns right from wrong. This inner moral compass was something altogether new on earth, setting man apart from the animal creation.

Furthermore, mankind was given qualities of wisdom and power that elevate him far above the creatures of the field. Job acknowledges Jehovah as *"perfect in knowledge"* and of *"almighty power"* (Job 37:23), while the apostle Paul speaks of God as *"the only wise God"* (Rom. 16:27). Though man's share in these qualities is limited, it is nevertheless a reflection of the One who granted them. Unlike the animals, man can reason, reflect, and appreciate the beauty of creation. He delights in the arts, in music, in speech, and in the exploration of knowledge—capacities that echo the mind and heart of his Maker.

Most significantly, man alone was given the gift of spirituality. Unlike animals, man can come to know his Creator, pray to Him, and cultivate a personal relationship with Him. This capacity for worship and communion marks the highest expression of what it means to be created in God's image. Through this, humans are able to reflect divine qualities not only inwardly but outwardly, in their conduct and relationships with one another.

Thus, when Genesis records that "God created man in his image" (Gen. 1:27), it does not mean in outward appearance but in inward qualities—qualities that make us capable of reflecting the very nature of God in love, justice, wisdom, and spirituality. This act of creation was purposeful and deliberate, taking place in 4026 B.C.E., as the Bible's chronology indicates—not as the result of countless ages of blind evolution. In this way, the creation of man stands as a testimony not only to God's power but to His intention that mankind should serve as His image-bearers on earth, capable of knowing Him and reflecting His glory.

Some have speculated that each creative "day" might have been 7,000 years, and that multiplied by seven days would equal 49,000 years. But that figure is an assumption, not a revelation. Scripture never gives such a calculation. What the Bible does emphasize is not the *length* of the days, but their *purpose*.

For example, in Genesis the word "day" (Hebrew: *yōm*) is used in more than one way. It can mean a literal 24-hour period, or it can refer to an extended, undefined span of time. In fact, even after describing the six days of creation, Genesis 2:4 summarizes the entire process as occurring "in the day that Jehovah God made earth and heaven." Clearly, here "day" means an era, not a single sunrise-to-sunset cycle.

So, while some traditions assign a set number of years to each creative day, the Bible itself leaves the duration open-ended. The inspired record directs us not to focus on the exact *length* of each period, but to recognize the *divine work* accomplished during it.

Day 1 – Light and Darkness (Genesis 1:3-5)

- God said: *"Let there be light."* Light began to reach the surface of the Earth, likely as the atmosphere cleared.
- This wasn't the creation of light itself (since the universe already existed), but rather the organization of light in relation to Earth.
- A distinction was made between **day** and **night**, setting the rhythm of time for Earth.

First Creative Day: The establishment of light and the separation of light from darkness (Ge 1:3–5) can be seen in the geological record as the earliest processes that made Earth's physical environment suitable for life. This includes the formation of the planet's atmosphere and the differentiation of day and night cycles—though these were not literal 24-hour periods, they mark the emergence of order from primordial chaos. The "evening" of this day represents the period when conditions were forming but not yet fully apparent, while the "morning" corresponds to the clear manifestation of light as a functional reality.

Day 2 – Sky and Waters (Genesis 1:6-8)

- God formed an "expanse" (*raqia* ʽin Hebrew), which separated the waters above from the waters below.

- The atmosphere took shape — water vapor remained above in the form of clouds, while liquid water gathered on Earth's surface.
- This created the breathable **sky** and stable weather cycles essential for life.

Second Creative Day: The separation of the waters above from the waters below, creating the sky and seas (Ge 1:6–8), aligns with the geological development of the hydrosphere and atmosphere. Water began to cycle, cloud formations emerged, and the foundations of Earth's climate systems were established. Geological evidence, such as ancient sedimentary layers and isotopic records, points to these processes occurring over vast spans of time. The "morning" of this day marks the visible organization of these waters into a functional system.

Day 3 – Land and Vegetation (Genesis 1:9-13)

- Waters were gathered into seas, and dry land appeared.
- Earth's continents and terrain began to stabilize.
- God then caused **vegetation** to sprout: grasses, seed-bearing plants, and fruit trees.
- This set up the foundation of Earth's food chain, providing nourishment for future creatures.

Third Creative Day: The emergence of dry land and the appearance of vegetation (Ge 1:9–13) corresponds to the geological formation of continents, mountain ranges, and fertile soil layers capable of supporting plant life. Fossilized plants and sedimentary deposits reveal the gradual establishment of terrestrial ecosystems. Here, the "evening" signifies the preparatory processes of soil formation and land emergence, while the "morning" represents the full manifestation of functional habitats for vegetation.

Day 4 – Luminaries Appointed (Genesis 1:14-19)

- The **sun, moon, and stars** became clearly visible from Earth's surface.
- They were appointed to serve as "signs, seasons, days, and years" — in other words, marking the passage of time.
- This also allowed mankind (later on) to use the heavens for calendars, navigation, and planting cycles.

Fourth Creative Day: The creation of the sun, moon, and stars (Ge 1:14–19) can be seen as the establishment of the celestial bodies' roles in regulating Earth's cycles, including seasonal and day-night rhythms. While the physical stars already existed, this day highlights their functional assignment in marking time and sustaining life on Earth. Geological and astronomical evidence suggests that Earth's orientation and rotation stabilized during this period, enabling predictable climatic and environmental patterns.

Day 5 – Sea Creatures and Flying Creatures (Genesis 1:20-23)

- Life in the **waters** was brought forth in abundance — fish, great sea creatures, and all aquatic life.
- Birds and other winged creatures began to populate the skies.
- God blessed them, telling them to "be fruitful and multiply."

Fifth Creative Day: The creation of sea creatures and birds (Ge 1:20–23) reflects the diversification of life in the oceans and the emergence of aerial species. Fossil records show a progressive development of marine life, followed by the appearance of early birds and other airborne organisms. The "morning" of this day corresponds to ecosystems achieving functional balance, with species fulfilling their intended roles.

Day 6 – Land Animals and Humankind (Genesis 1:24-31)

- Animals appeared on dry land: livestock, wild beasts, and creeping things.
- Finally, the climax of creation: **humankind.**
- Man and woman were made in God's image — not in physical form, but in qualities like reason, morality, creativity, and the capacity for spirituality.
- Humanity was given responsibility to care for the Earth and all living things.

Sixth Creative Day: The creation of land animals and humans (Ge 1:24–31) aligns with the final stages of terrestrial life development. Fossil evidence documents the emergence of mammals and, eventually, anatomically modern humans. This day encompasses the full functioning of ecosystems, with humans as stewards, uniquely made in the image of God. The "morning" signifies the clear realization of God's purpose in preparing the Earth for human habitation.

Day 7 – God's Rest (Genesis 2:1-3)

- God ceased from creative activity.
- This doesn't mean He became inactive — rather, He entered into a period of **rest** from His creative activity related to the Earth.
- Hebrews 4:3-10 shows that this seventh "day" continues, meaning it is ongoing.

Thinking outside of the box: Creative Days

Concluding the review of accomplishments on each of the six days of creative activity is the recurring statement: *"And there came to be evening and there came to be morning"—first day, second day, third day,* and so forth (Ge 1:5, 8, 13, 19, 23, 31). It is clear, however, that the length of each creative day far exceeded a literal 24-hour period. Therefore, this expression should not be interpreted as referring to a literal night and day but as figurative language.

During the "evening" phase of each creative period, things would remain indistinct or not fully apparent. By the "morning," however, God's purpose for that day would be clearly manifested and fully accomplished. While the divine purpose of each day was always fully known to God, angelic observers or other beings would only perceive the outcome as it became evident in the "morning."

There is no indication that all Creative Days were equal in length. Each day would last as long as necessary to achieve God's intended purpose. Some days could be longer, others shorter, depending on the complexity or scope of the creative activity involved. Moreover, there is reason to believe that certain Creative Days may have overlapped, rather than occurring in strict sequence.

With this perspective, it becomes possible to explore a potential relationship between the sequence of Creative Days and geological findings, offering a framework to align observed natural processes with the order of God's creative work.

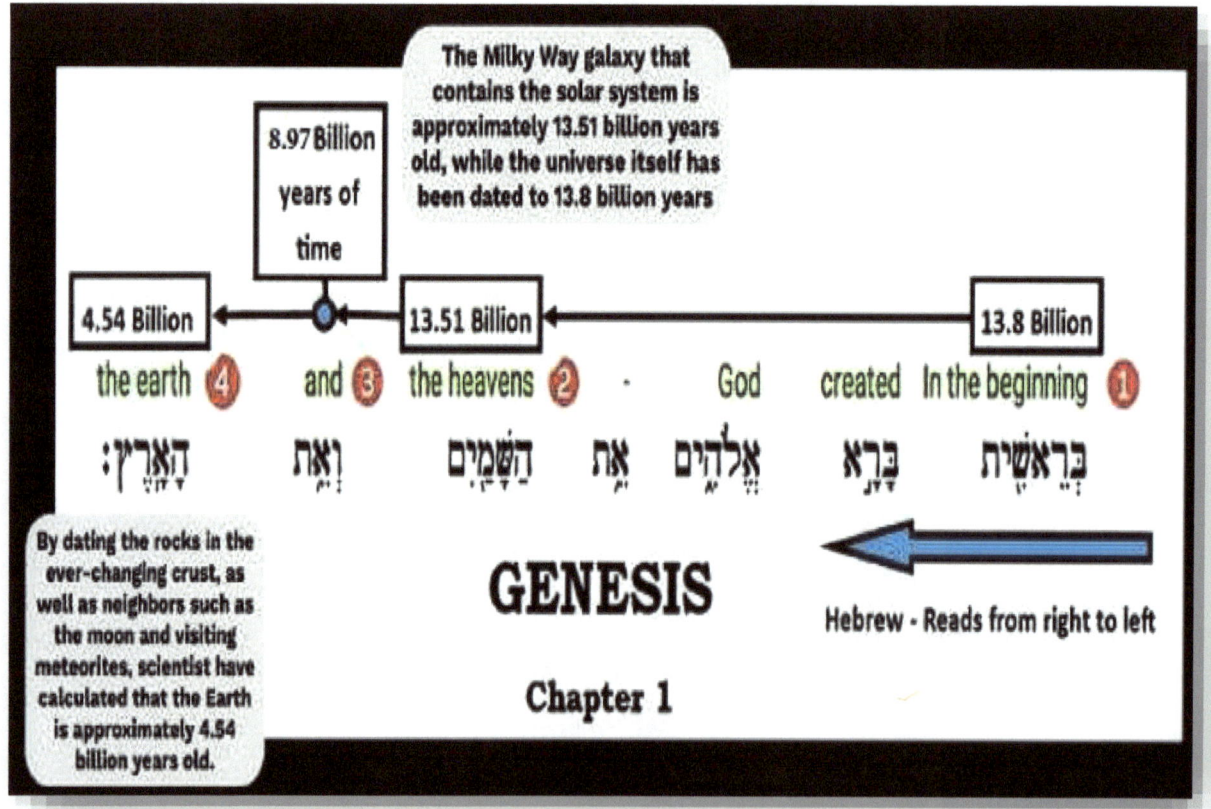

Now, let's look at a chart with Geological dates and how the Creative Days 1 through 6 of Genesis.

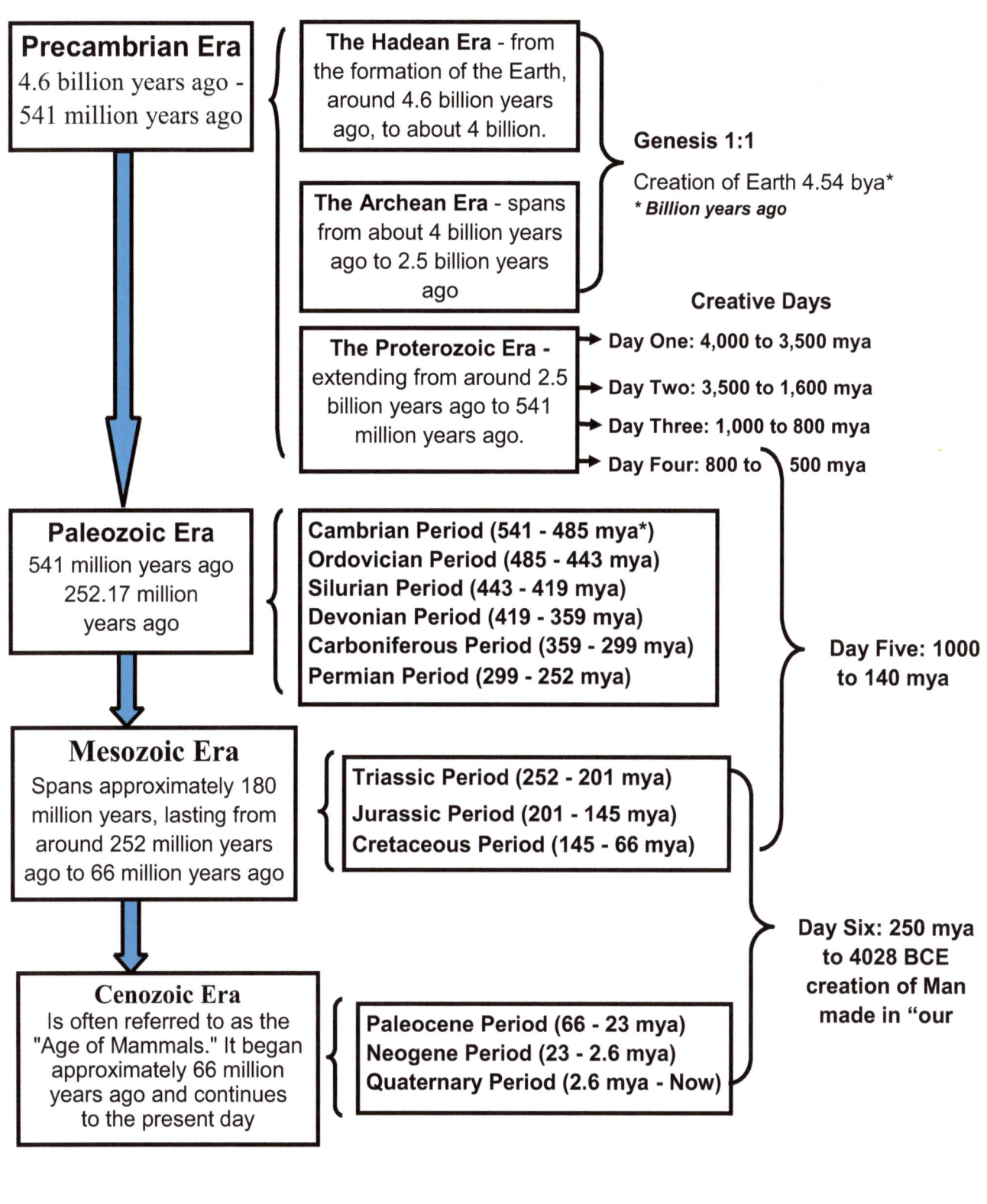

Precambrian Era
4.6 billion years ago -
541 million years ago

The Hadean Era - from the formation of the Earth, around 4.6 billion years ago, to about 4 billion.

The Archean Era - spans from about 4 billion years ago to 2.5 billion years ago

The Proterozoic Era - extending from around 2.5 billion years ago to 541 million years ago.

Genesis 1:1

Creation of Earth 4.54 bya*
Billion years ago

Creative Days

Day One: 4,000 to 3,500 mya

Day Two: 3,500 to 1,600 mya

Day Three: 1,000 to 800 mya

Day Four: 800 to   500 mya

Paleozoic Era
541 million years ago
252.17 million
years ago

Cambrian Period (541 - 485 mya*)
Ordovician Period (485 - 443 mya)
Silurian Period (443 - 419 mya)
Devonian Period (419 - 359 mya)
Carboniferous Period (359 - 299 mya)
Permian Period (299 - 252 mya)

Day Five: 1000
to 140 mya

Mesozoic Era
Spans approximately 180 million years, lasting from around 252 million years ago to 66 million years ago

Triassic Period (252 - 201 mya)
Jurassic Period (201 - 145 mya)
Cretaceous Period (145 - 66 mya)

Day Six: 250 mya
to 4028 BCE
creation of Man
made in "our

Cenozoic Era
Is often referred to as the "Age of Mammals." It began approximately 66 million years ago and continues to the present day

Paleocene Period (66 - 23 mya)
Neogene Period (23 - 2.6 mya)
Quaternary Period (2.6 mya - Now)

*mya = one million years ago

References: Bing com, wol.jw.org. bibleview org.

Biblical Days Five & Six with matching Geo Dates

Creative Day Five

Aquatic souls and flying creatures
Genesis 1:20-23

Genesis 1:20
Then God said: "Let the waters swarm with living creatures and let flying creatures fly above the earth across the expanse of the heavens."

Next we have listed Era with sub sections that make up the Fifth Day of Creation
Paleozoic Era:

Cambrian explosion - During this time, **arthropods**, **mollusks**, and other marine organisms proliferated in the oceans.

Silurian period, life transitioned from water to land
Leafless vascular plants (psilophytes) and **invertebrate animals** (such as centipede-like arthropods) established themselves on land.

Devonian and Permian period witnessed the rise of **fish** and the emergence of the first **amphibians***.

---

*Ichthyosaurs (250 mya)

Creative Day Six

Ge 1:24-31- Land animals; man

Genesis 1:24
Then God said: "Let the earth bring forth living creatures according to their kinds, domestic animals and creeping animals and wild animals of the earth according to their kinds." And it was so.

Thinking outside of the box: Creative Days

Now we will list the Epochs with sub sections that make up part of the Fifth and Sixth Day of Creation

Mesozoic Era: The end of the Mesozoic witnessed one of the most significant events: the **mass extinction** that led to the demise of the dinosaurs and the rise of mammals and birds1.

- **Triassic Period**: This period marked the rise of **archosaurs**, a group that includes **dinosaurs, crocodiles,** and **birds.**
- **Jurassic Period**: age of dinosaurs Sauropods (long-necked herbivorous dinosaurs), **theropods** (carnivorous dinosaurs), and **stegosaurs** (plated dinosaurs) flourished.

Triceratops

- **Cretaceous Period**: Tyrannosaurus rex, Triceratops, and **Velociraptors** were some of the iconic dinosaurs of this era.

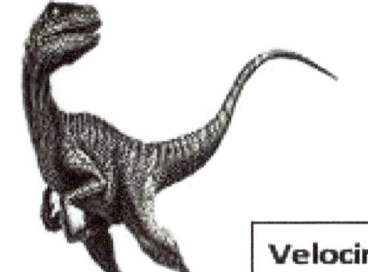

Velociraptor

Cenozoic Era: often called the **"Age of Mammals"** because mammals dominated the land during this time. The largest land animals were mammals, and they diversified into various forms.

  Paleogene Period: (66 million years ago - 23.03 million years ago)

Extinction of Dinosaurs: At the dawn of the Paleogene, non-avian dinosaurs, *pterosaurs, and giant marine reptiles were absent.

With the above information I do believe there's no problem in blending together Creative days with Geo/ Scientific dates.

In conclusion, what am I asking each side of the discussion to do? Change your thinking, look outside the box and view things from a different angle

*Pterosaurs

Thinking outside of the box: Creative Days

Do you see any conflict of interest? None at all. And that's because, when we broaden our perspective beyond the conventional frameworks of geological and cosmic dating, no real conflict exists.

From a Biblical standpoint, there is no requirement that each Creative Day be of equal length. Each day could last precisely as long as necessary to fulfill God's intended purpose. Since God exists outside of time, there is no urgency—some Creative Days may have been extraordinarily long, while others may have been comparatively brief, depending entirely on their objectives.

Furthermore, evidence allows for the possibility that some Creative Days overlapped. Neither Biblical principles nor the natural laws observed in geology or science contradict such a scenario. In this light, the chronology of creation is not a rigid timeline but a purposeful unfolding of God's plan.

When it comes to thinking outside the box, I like to reflect on the thoughts of Bettina J. Casad, Assistant Professor in the Department of Psychological Sciences at the University of Missouri– St. Louis. Her contributions to SAGE Publications' Encyclopedia of Social Psychology (2007) formed the basis of her contributions to Britannica. She said this: "Confirmation Bias is the tendency to process information by looking for, or interpreting, information that is consistent with one's existing beliefs. This biased approach to decision-making is largely unintentional and often results in ignoring inconsistent information. Existing beliefs can include one's expectations in a given situation and predictions about a particular outcome. People are especially likely to process information to support their own beliefs when the issue is highly important or self-relevant."

So then, everyone needs to open their mind to a different point of view.

> "To raise new questions, new possibilities, to regard old problems from a new angle, requires creative imagination and marks real advance in science."
> Albert Einstein[14]

Rethinking Creation: Bridging Science and Faith

The way we understand the world—both scientifically and spiritually—often depends on the assumptions we bring to it. For science, this means being willing to explore the possibility that a Creator exists. For religion, it requires openness to insights from geology, cosmology, and evolutionary study. Neither perspective has to negate the other; in fact, both can complement one another when we approach them with humility and curiosity.

From a scientific standpoint, there is reason to question the idea of a continuous, fully traceable evolutionary line. The so-called "missing link" between Neanderthals and humanity remains elusive, leaving a gap of roughly 30,000 years between early hominids and humans described as

Thinking outside of the box: Creative Days

"made in God's image." Rather than viewing evolution as a purely random process, it may be more accurate to consider it as part of a larger, ongoing creative pattern—one that aligns with a purposeful design behind life itself.

Religious thought, likewise, can benefit from a shift in perspective. The Hominoidea—primates including Neanderthals—might represent an earlier stage of humanity's development during the sixth Creative Day. At the same time, the geological and astronomical dating of the universe is not inherently in conflict with Scripture; the six Creative Days need not have been uniform in length and could even have overlapped. By embracing these possibilities, religious traditions can maintain fidelity to Scripture while also integrating the evidence science provides.

Ultimately, bridging science and faith requires a willingness to rethink old assumptions. When approached with openness, neither discipline loses its integrity; instead, each gains depth, allowing for a richer understanding of creation, purpose, and the extraordinary unfolding of life.

"The scientific mind does not so much provide the right answers
as ask the right questions."
Claude Levi-Strauss14

"We cannot solve our problems with the same thinking we
used when we created them."
Albert Einstein14

Thinking outside of the box: Creative Days

Let's cite some examples of harmony between the Biblical record and Scientific understanding?

2nd Law of Thermodynamics

First of all, you need to realize that the Bible is logical and has been the basis for many scientific discoveries. I'm thinking about the Second Law of Thermodynamics. In the 19th century, scientist William Thomson (also known as Lord Kelvin) is credited with discovering the Second Law of Thermodynamics, which explains why, over time, *natural systems tend to decay and break down*. One factor that inspired him to uncover this law was a careful study of the Bible book of Psalm 102:25-26[1,] which says:

Lord Kelvin

Psalm 102:25 "Long ago you laid the foundations of the earth, and *the heavens* are the work of your hands. 26 They will perish, but you will remain. *Just like a garment, they will all wear out.* Just like clothing, you will replace them; they will pass away." [NWT]

The key was. "*The heavens … Just like a garment they will all wear out.*" This scripture was written 2,284 years ago, and the Second Law of Thermodynamics was discovered in 1824 C.E.

Thinking outside of the box: Creative Days

A Flat Earth:

Here is an additional scientific fact from the Bible. For years, it was thought that the Earth was flat. Another scripture is found at Job 26:7 (written 1473 B.C.E.)

Isaiah 40:22a, "There is One who dwells above *the circle of the earth*, and its inhabitants are like grasshoppers." [NWT]

It has been 3,434 years since the first astronaut orbited the Earth and acknowledged that the Earth

hangs in space with no visible support. The knowledge of a round Earth was gradually adopted throughout the Old World during Late Antiquity and the Middle Ages, from the 5th to the late 15th century (400 to 1400 C.E.).

Job 26:7 "He stretches out the northern sky over empty space*, [GOD]* ***hangs the earth upon nothing;"*** [NWT]

The ancients believed that the Earth was supported by giant animals, elephants, turtles, and a snake as it floated through the Cosmos (heavens).

Thinking outside of the box: Creative Days

Earth's Water *(hydrologic)* Cycle:

It was not until the 1800s that the Earth's water cycle was relatively well understood. However, it was mentioned 3,273 years ago in the Bible.

Job 36:27-28
He (God) **draws up the drops of water**; They **condense into rain from his mist**; [28] Then the clouds pour it down; They **shower down upon mankind**". [NWT] Written 1473 BCE.

2 *Condense into rain from his mist.*

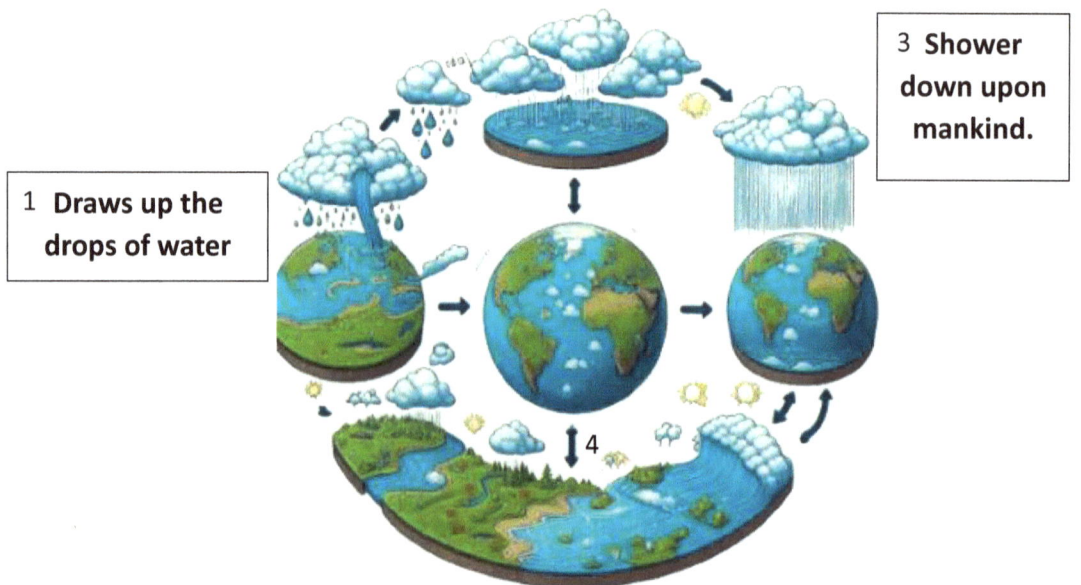

1 **Draws up the drops of water**

3 **Shower down upon mankind.**

Amos 9:6 – Speaks of God "calling for the waters of the sea and pouring them out upon the surface of the earth."

1. Evaporation The cycle begins when the sun heats bodies of water — oceans, lakes, and rivers. That heat energy causes water to change into vapor and rise into the atmosphere. Plants also release water vapor through their leaves in a process called **transpiration**.

2. Condensation As the water vapor rises higher into cooler air, it condenses into tiny droplets, forming clouds. This process is what makes clouds visible in the sky — they're actually made of countless tiny water droplets or ice crystals.

3. Precipitation When those droplets combine and grow heavy enough, gravity pulls them back to Earth in the form of rain, snow, sleet, or hail. This returns fresh water to the surface.

4. Collection The precipitation collects in rivers, lakes, soil, and underground reservoirs called **aquifers**. Most of it eventually flows back to the oceans, which serve as the cycle's main "storage tank."

Thinking outside of the box: Creative Days

Cloud Physics - Branch of Atmospheric Science:

Luke Howard (1802) is often referred to as "The Godfather of Clouds," the one who named the basic cloud types, and the "father of meteorology".

Clouds appear to float in the air because the water droplets and ice crystals that make up the clouds are tiny and have low fall velocities. They are also carried by upward vertical motions or updrafts in the atmosphere, which counteract the effects of gravity.

However, Job 26:8 was written in 1473 B.C.E., which was 3,275 years before the concept of Atmospheric Science was first developed.

Job 26:8 He (God) wraps up the waters in his clouds, so that the clouds *do not burst under their weight*. [NWT]*

Luke Howard
1802

1. Puffy and white
Cumulus cloud

2. Stratocumulus cloud

3. Cumulonimbus or
thunderstorm cloud

1. Puffy and white clouds - Can hold up to .05-2 million lbs. of water.

2. Stratocumulus cloud - Can hold 2-3 million lbs. of water.

3. Cumulonimbus or thunderstorm cloud - Can hold up to 100 million lbs. of water. A 100 million lbs. of water = 8,986,819 square feet.

Several man-made reservoirs and lakes are pretty large and could potentially have an area of around 100 million lbs. of water. Some examples include: Lake Mead (USA); Lake Volta (Ghana); Great Salt Lake (USA)

> * For centuries, air was thought to be weightless. But in the 1600s, scientists discovered that air actually has mass and exerts pressure — the basis of barometers and weather forecasting. Job's words anticipated this reality by thousands of years.

Thinking outside of the box: Creative Days

Dark Matter and Dark Energy:

Dark matter, a ***mysterious substance that accounts for approximately 27% of the universe's mass-energy content, is akin to an unseen fine gauze***, according to current cosmological models. Despite its prevalence, dark matter does not emit, absorb, or reflect electromagnetic radiation, making it invisible and detectable only through its gravitational effects on visible matter.

The Bible first spoke of this Dark Matter as a **fine gauze at Isaiah 40:22b over 2,665 years ago.

NWT Study Bible - Isaiah 40:22b He (God) **is *stretching out the heavens like a fine** gauze*, and he *spreads them out like a tent* to dwell in.”** Written 732 B.C.E.

Here's a breakdown of some key aspects of dark matter:

Gravitational Effects: Dark matter's presence is inferred primarily through its gravitational effects on galaxies and galaxy clusters. Observations of galaxy rotation curves, gravitational lensing, and the large-scale structure of the universe all suggest the presence of vast amounts of unseen matter.

Dark Energy: Dark matter should not be confused with dark energy, another mysterious component of the universe that constitutes about 68% of its mass-energy content. While dark matter acts as a gravitational glue, dark energy is responsible for the universe's accelerated expansion.

Fritz Zwicky

Composition: In the standard cosmological model (lambda-CDM), the universe's mass-energy content consists of:

5% ordinary matter: The stuff we see, including stars, planets, and galaxies.

26.8% dark matter: This mysterious substance constitutes the majority of a galaxy's mass.

68.2% dark energy: A form of energy responsible for the universe's accelerated expansion.

Originally known as the "missing mass," dark matter's existence was first inferred by Swiss American astronomer Fritz Zwicky, who in 1933 was the first to use the virial theorem to postulate the existence of unseen dark matter. Despite decades of research, many questions about dark matter remain unanswered. Its exact composition, properties, and interactions with ordinary matter are still subjects of intense study and debate in the fields of cosmology, particle physics, and astrophysics.

Thinking outside of the box: Creative Days

Gravity:

Gravity (cords) is a natural phenomenon by which all things with mass or energy, including planets, stars, galaxies, and even light, are attracted to (or gravitate toward) one another. The gravitational attraction of the original gaseous matter present in the Universe caused it to begin coalescing and forming stars. It caused the stars to group into galaxies, so gravity is responsible for many of the larger-scale structures in the Universe.

Sir Isaac Newton published a comprehensive theory of gravity in 1687 C.E., 3,160 years after the book of Job was written.

The Bible alluded to the concept of gravity as far back as 1473 BCE in the book of Job.

Job 38: 31; 33
Can you tie fast ***the bonds* of the Ki'mah# constellation,*** or can you loosen ***the very cords* of the Ke'sil constellation?*** 33 Have you come to know ***the statutes of the heavens,*** or could you put its authority in the earth? [*Synonyms for bonds: cords, cement, knot, ligature, tie, fixes, adheres.]

KIMAH CONSTELLATION (Ki'mah)

This term is used at Job 9:9, 38:31, and Amos 5:8 to refer to a celestial constellation. It is usually considered to refer to the Pleiades, a star group comprising seven large stars and other smaller ones, enveloped in nebulous matter and situated approximately 380 light-years from the Sun.

Sir Isaac Newton was a brilliant English mathematician, physicist, astronomer, and author who is widely regarded as one of the most influential scientists of all time. He was born on December 25, 1642, in Woolsthorpe, Lincolnshire, England, and died on March 20, 1727, in Kensington, London.

Newton made groundbreaking contributions to many areas of science, including mathematics, physics, and astronomy. Perhaps his most famous achievement was his formulation of the laws of motion and universal gravitation, which were laid out in his monumental work, "Philosophiæ Naturalis Principia Mathematica" ("Mathematical Principles of Natural Philosophy"), commonly known as the Principia, published in 1687.

Newton's influence extended beyond his scientific contributions; he also played a key role in the scientific revolution of the 17th century and the Enlightenment era

Sir Isaac Newton

that followed. He was knighted by Queen Anne in 1705 and served as President of the Royal Society from 1703 until he died in 1727.

Sir Isaac Newton was a deeply religious individual who held strong beliefs in God. His religious convictions played a significant role in his life and work. Newton was a devout Christian, and his writings indicate that he saw his scientific pursuits as a means of understanding God's creation.

Newton's theological views were complex and sometimes unconventional, as he delved into topics such as biblical prophecy and the nature of God. He wrote extensively on religious matters, producing more manuscripts on theology and alchemy than on physics and mathematics combined.

While Newton's scientific discoveries are often celebrated for their role in advancing natural philosophy and laying the groundwork for modern science, it's essential to recognize that he saw no conflict between his scientific pursuits and his religious beliefs. Instead, he viewed them as complementary aspects of understanding the universe.

Quarantine and Disease:

Leviticus 13:2 "If a man develops on his skin a swelling, a scab, or a blotch and it could become the disease of leprosy *4* … the priest will then quarantine the infected person for seven days.

Written about 1500 B.C.E

then shall isolate	the priest	-	the [one who has the] sore	seven	days
וְהִסְגִּיר	הַכֹּהֵן	אֶת־	הַנֶּגַע	שִׁבְעַת	יָמִים:
wə·his·ĝîr	hak·kō·hên	'et–	han·ne·ḡa'	šib·'at	yā·mîm

C.E. 549 (6th Century - 2,040 after the book of Leviticus was written) In the wake of one of history's most devastating epidemics of bubonic plague, the Byzantine emperor Justinian enacts a law meant to hinder and isolate people arriving from plague-infested regions. It wasn't until the Black Death of the 14th century, however, that Venice established the first formal system of quarantine, requiring ships to lie at anchor for 40 days before landing. ("Quarantine" comes from the Latin for forty.) - *BY P. TYSON – NOVA OCTOBER 2004.*

Justinian Plague 549 C.E.

Hurdles:

When we begin comparing the Genesis creation account with scientific perspectives on human origins and the universe, one of the first hurdles that comes up is **timeline and sequence**. Genesis presents a six-day creation, culminating with humans on the sixth day. Evolutionary theory, however, stretches across millions of years and describes a gradual unfolding of life from simple organisms to complex beings, including humans. While Genesis and science both place humans after the animals, the processes and timescales are vastly different, raising questions about how (or if) the two accounts can align.

Another major point of discussion revolves around **Adam and Eve**. For Christians, their direct creation by God is central—not only to the creation story but to doctrines of sin and redemption. Evolutionary theory doesn't allow for a single original couple; instead, it speaks of populations of early hominins gradually developing. The contrast between a single, divinely created pair and a slow evolutionary process of many is a significant theological and scientific tension.

Then there's the **fossil record**. Evolutionists argue that fossils reveal a gradual transition from ape-like ancestors to modern humans. Genesis doesn't reference Neanderthals or other hominin species, leaving room for debate about where these beings fit in—were they animals, proto-humans, or something else entirely? Some suggest they could be incorporated into the Genesis framework, while others see them as contradictory.

The way one approaches these questions often depends on **literal versus symbolic interpretation**. A literalist might insist on a young Earth and six actual 24-hour days, leaving little room for Geo-scientific timelines. Others take a symbolic approach, viewing the "days" as long epochs or poetic structures, which allows them to reconcile Genesis with scientific models.

The tension grows sharper with the concept of **common ancestry**. Evolution reduces humanity to just another branch of life's tree, while Genesis sets humans apart, uniquely made in God's image. Here lies one of the greatest divides: are we accidents of natural selection, or beings with a divine imprint and eternal value? Only the Genesis account affirms the latter with clarity. Science without God leaves us with randomness and chance. Genesis presents us with design, meaning, and human dignity. The hurdles are real, but they invite us to ask the deeper question: Which vision of origins gives us not only an account of how we came to be, but also a reason why we exist at all?

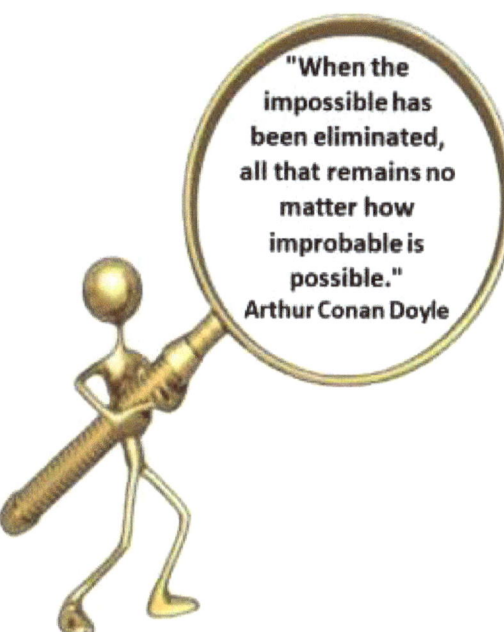

"When the impossible has been eliminated, all that remains no matter how improbable is possible."
Arthur Conan Doyle

In the beginning, God created the heavens and the earth...

BANG!

REFERENCES:

1 New World Translation of the Bible

2 New York, 1980; pg. 219

3 nationalgeographic.com

4 nps.gov

5 uwlax.edu

6 thecanadianencyclopedia.ca 7 Scientificamerican.com

8 sljinstitute.net

9 academia.edu

10 britannica.com

11 https://www.britannica.com/topic/hominin

12 Animal Sciences. Encyclopedia.com. 10 Feb. 2018 13 g73 10/22 pp.18-21 14

www.brainyquote.com/quotes/albert_einstein_121993

ADDITIONAL SOURCES:

All Arthur Conan Doyle quotes: /www.brainyquote.com/quotes/arthur_conan_doyle_398910

J. Green. The Interlinear Hebrew/ Greek English Bible - Vol One.

Bible: King James Version, NWT Study version

Translation Committee.. WTB & Tract Society Inc. of Brooklyn, N.Y. Bible, Greek Interlinear

Christine Hobson. The World of the Pharaohs, Thames & Hudson Inc - 1990

M Brunson. A Dictionary of Ancient Egypt, Oxford University Press , N.Y., 1991

Ian Shaw. Oxford History of Ancient Egypt. Oxford University Press, Butler & Tanner, Ltd., -Great Britain -2000/ 2002/ 2003

Translation Committee. NWT Of Holy Scriptures - 1984; WBT & Tract Society of Pa.

Carl Sagan. Cosmos - (New York, 1980), p. 29

Peter A. Clayton. Chronicle of the Pharaohs, Thames and Hudson Inc., London. - 2006

Translation Committee. WTB & Tract Society Inc. of Brooklyn, N.Y. Bible: American Standard Version

www.merriamwebster.com

Animal Sciences. Encyclopedia.com. 10 Feb. 2018

Publishing Committee . it-1 p. S45 Creation - WTB & Tract Society Inc. of Brooklyn, N.Y

By J. Wash Watts. W. B. Eerdmans, Grand Rapids, Michigan, 1963. A Distinctive Translation of Genesis

Translation Committee. WTB & Tract Society Inc. of Brooklyn, N.Y. Bible: Byington Version

Publishing Committee. si p. 14 par. 9 Bible Book No. ,-Genesis; WTB& Tract Society of Pa.

www.scifaets.net/timelines/

Translation Committee. WTB & Tract Society Inc. of Brooklyn, N.Y. Reference Bible

Publishing Committee. Was life Created? Science and the Genesis account, pg. 26-27; WTB& Tract Society of Pa.

Bio.libretexts.org

Usgs.gov

thefamouspeople.com

Thinking outside of the box: Creative Days

Carl Sagan was an American astronomer, cosmologist, astrophysicist, astrobiologist, author, science communicator, and science popularizer. Born on November 9, 1934, in Brooklyn, New York, he became one of the most prominent figures in the popularization of science in the 20th century.

Sagan earned his bachelor's and master's degrees in physics from the University of Chicago and obtained his Ph.D. in astronomy and astrophysics from the University of Chicago in 1960. He was a prolific researcher, contributing to fields such as planetary science, astronomy, and astrobiology. He worked on various NASA projects, including the Mariner, Viking, Voyager, and Galileo missions.

One of Sagan's most significant contributions to science was his work on the possibility of extraterrestrial life. He was a key figure in the development of the SETI(Search for Extraterrestrial Intelligence) program and helped popularize the idea of searching for life beyond Earth. Sagan was also a talented writer and communicator. He authored numerous scientific papers and popular science books, including "Cosmos," which accompanied his Emmy and Peabody Award-winning television series of the same name. "Cosmos" explored a wide range of scientific topics, from the origins of the universe to the search for extraterrestrial life, and it became one of the most widely watched series in the history of American public television.

Beyond his scientific work, Sagan was known for his advocacy of critical thinking, skepticism, and the scientific method. He often spoke out against pseudoscience and irrational thinking, emphasizing the importance of evidence-based reasoning.

Sagan was also an advocate for nuclear disarmament, environmentalism, and the responsible use of technology. He served as a consultant and adviser to various governmental and non-governmental organizations, providing insights into scientific matters and public policy.

Carl Sagan passed away on December 20, 1996, but his legacy continues to inspire scientists, educators, and science enthusiasts around the world. His contributions to astronomy, planetary science, and science communication have left an indelible mark on the field of science and society as a whole.

Thinking outside of the box: Creative Days

William Thomson, 1st Baron Kelvin, commonly known simply as Lord Kelvin, was a prominent physicist, mathematician, and engineer of the 19th century. Born on June 26, 1824, in Belfast, Ireland (now Northern Ireland), he made significant contributions to various fields of science and technology.

Early Life and Education: William Thomson was the fourth child in a family of seven. He showed early aptitude for mathematics and science. He attended the University of Glasgow at the age of 10 and later continued his studies at Peterhouse, Cambridge. Kelvin held the position of **Professor of Natural Philosophy** at the **University of Glasgow** for an impressive **53 years**.

Thermodynamics and Kelvin Scale: Thermodynamics: Kelvin significantly contributed to the formulation of the first and second laws of thermodynamics. His work laid the foundation for our understanding of energy transfer and heat.

The 2nd Law Of Thermodynamics explained in an easier-to-understand way:

The second law of thermodynamics is all about how things change over time. Basically, it says that in any natural process, the total amount of disorder, or entropy, in the universe tends to increase.

Imagine you have a neat and organized room. According to the second law, if you leave it alone, it's more likely to get messy over time than to tidy up by itself. That's because disorder naturally increases.

In simpler terms, things left on their own tend to become more chaotic. This law helps us understand why heat flows from hot to cold objects, why machines aren't 100% efficient, and why time only moves in one direction – from past to future.

Absolute Temperature Scale: He invented the international system of absolute temperature, which is measured in **kelvins (K)** in his honor. The Kelvin scale starts from absolute zero, the point at which molecular motion ceases. The Kelvin scale is now the standard unit of temperature measurement in scientific contexts.

Electromagnetism: Kelvin made important contributions to electromagnetic theory, including the electromagnetic theory of light. He formulated Thomson's theorem in electrostatics and conducted research on the behavior of electrical circuits. He also made significant contributions to mathematical physics, particularly in the field of potential theory.

Telegraph Cable: Thomson played a crucial role in the development of the transatlantic telegraph cable. He worked on the problem of signal distortion in long submarine cables and developed techniques to improve signal transmission. His expertise was instrumental in the successful laying of the first transatlantic telegraph cable in 1858.

Engineering and Navigation: Kelvin made notable contributions to engineering, particularly in the field of maritime navigation. He developed accurate compasses and instruments for measuring depths, which greatly improved navigation and safety at sea.

Titles and Honors: In recognition of his contributions to science and engineering, Thomson was knighted by Queen Victoria in 1866 and later elevated to the peerage as Baron Kelvin of Largs in 1892. The unit of temperature, the Kelvin, was named in his honor.

Legacy: Lord Kelvin's work laid the foundation for many areas of modern physics and engineering. His contributions to thermodynamics, electromagnetism, and navigation have had a lasting impact on science and technology. He is remembered as one of the most influential scientists of the 19th century.

Luke Howard (1772–1864) He was a British manufacturing chemist, amateur meteorologist, and a significant figure in the scientific community. His contributions spanned various fields, but his most lasting impact lies in meteorology. Here are key aspects of his life and work:

Nomenclature System for Clouds: In 1802, Howard proposed a systematic classification and naming system for clouds during a presentation to the Askesian Society. His work categorized clouds into three main types: **cumulus**, **stratus**, and **cirrus**. Due to this groundbreaking contribution, Howard is often referred to as "The Godfather of Clouds" and "The Father of Meteorology."

Personal Life: Born in London in 1772, Howard came from a Quaker family. He attended a Quaker grammar school in Oxfordshire. Howard married and had two sons, Robert and John Eliot, who later took over their father's chemical manufacturing business. Despite being a Quaker, he left the Society in 1825 due to a dispute.

Career and Scientific Work: Howard worked as a pharmacist and chemist. He partnered with fellow Quaker William Allen to form the pharmaceutical company **Allen and Howard**. His factory in Plaistow, east of London, produced industrial chemicals and pharmaceuticals. Howard's comprehensive weather recordings in London from 1801 to 1841 transformed meteorology. He was elected a Fellow of the Royal Society in 1821.

Legacy: Howard's cloud classification system remains in international use today. His passion for meteorology earned him the title "father of meteorology." His observations and writings influenced science, art, and culture.

Luke Howard's dedication to understanding the skies continues to shape our understanding of weather and clouds.

Fritz Zwicky (1898–1974) was a Swiss astronomer who made significant contributions to theoretical and observational astronomy during his career at the California Institute of Technology (Caltech) in the United States. Here are some key points about his life and work:

Dark Matter Hypothesis: In 1933, Zwicky was the first to use the **virial theorem** to postulate the existence of **unseen dark matter** in the universe. He described this mysterious matter as "dunkle Materie" (German for dark matter).

Early Life and Education: Born in Varna, Bulgaria, to a Swiss father and a Czech mother, Zwicky had a diverse cultural background. He studied mathematics and experimental physics at the Swiss Federal Polytechnic (now ETH Zurich) and earned his doctorate with a thesis on heteropolar crystals.

Career and Contributions: Zwicky emigrated to the United States in 1925 and worked with physicist Robert Millikan at Caltech. His groundbreaking work included:

> Coining the term **"supernova"** while fostering the concept of **neutron stars**.
> Making important contributions to our understanding of galaxies as **gravitational lenses**.
> Serving as a research director/consultant for Aerojet Engineering Corporation.
> Being appointed Professor of Astronomy at Caltech in 1942.

Legacy and Honors: Zwicky's impact extends beyond academia, influencing science, art, and culture. He received awards such as the **President's Medal of Freedom** and the **Gold Medal of the Royal Astronomical Society**.

Zwicky's curiosity and innovative thinking continue to inspire astronomers and scientists worldwide.

Thinking outside of the box: Creative Days

Arthur Conan Doyle was a Scottish writer best known for creating the fictional detective Sherlock Holmes. Here's a comprehensive overview of his life and achievements:

Early Life: Arthur Ignatius Conan Doyle was born on May 22, 1859, in Edinburgh, Scotland. He was the eldest son of Charles Altamont Doyle, a civil servant and chronic alcoholic, and Mary Doyle. His early education took place at Jesuit schools, followed by further studies at Stonyhurst College.

Medical Career: Conan Doyle studied medicine at the University of Edinburgh and received his medical degree in 1881. He worked as a ship's surgeon on a whaling voyage to the Arctic and later established a medical practice in Southsea, England. Despite his medical career, he struggled to establish a successful practice and began writing to supplement his income.

Creation of Sherlock Holmes: Conan Doyle's most famous literary creation, Sherlock Holmes, made his debut in the novel "A Study in Scarlet," published in 1887. Holmes, a brilliant detective with remarkable powers of observation and deduction, quickly became a literary sensation. Conan Doyle went on to write three more novels and 56 short stories featuring Holmes and his loyal companion, Dr. John Watson.

Popularity and Impact: Sherlock Holmes became one of the most enduring and iconic characters in detective fiction, influencing countless other authors and popular culture. The stories featuring Holmes and Watson are celebrated for their intricate plots, memorable characters, and ingenious mysteries.

Other Literary Works: Despite Holmes's overwhelming popularity, Conan Doyle wrote prolifically in other genres as well. He authored historical novels, science fiction, fantasy, and adventure stories. His works include "The Lost World," featuring the character Professor Challenger, and numerous standalone novels and short stories.

Thinking outside of the box: Creative Days

Later Life and Honors: Conan Doyle was knighted in 1902 for his support of the British Empire during the Boer War. He continued to write and lecture throughout his life, although he increasingly focused on spiritualism in his later years. Conan Doyle passed away on July 7, 1930, in Crowborough, East Sussex, England, leaving behind a rich literary legacy that continues to captivate readers worldwide.

Arthur Conan Doyle's contributions to literature, particularly through the creation of Sherlock Holmes, have left an indelible mark on popular culture and established him as one of the greatest writers of detective fiction in history.

Thinking outside of the box: Creative Days

Albert Einstein was one of the most influential physicists of the 20th century, best known for his theory of relativity and his contributions to the development of quantum mechanics. Here's a comprehensive overview of his life and achievements:

Albert Einstein was born on March 14, 1879, in Ulm, in the Kingdom of Württemberg in the German Empire (now in Germany). He grew up in a secular Jewish family and showed an early interest in mathematics and science. Einstein attended the Swiss Federal Institute of Technology in Zurich, Switzerland, where he studied physics and mathematics. After graduating in 1900, he struggled to find academic employment and worked for several years as a patent clerk in Bern, Switzerland.

Annus Mirabillis Papers: In 1905, Einstein published four groundbreaking papers that revolutionized physics and established his reputation as a leading scientist. These papers introduced his theory of special relativity, which transformed our understanding of space, time, and energy, as well as his explanation of the photoelectric effect, which laid the groundwork for quantum theory.

Building on his work on special relativity, Einstein developed his theory of general relativity, which he published in 1915. General relativity explains how gravity arises from the curvature of spacetime, which is caused by the presence of mass and energy. It has profound implications for our understanding of the universe, predicting phenomena such as the bending of light around massive objects and the existence of black holes.

In 1921, Einstein was awarded the Nobel Prize in Physics for his explanation of the photoelectric effect, rather than for his work on relativity. This was because the photoelectric effect had more immediate experimental verification and practical applications.

Einstein emigrated to the United States in 1933 to escape the rise of Nazism in Germany. He settled in Princeton, New Jersey, where he joined the Institute for Advanced Study. In addition to his scientific work, Einstein was an outspoken advocate for pacifism, civil rights, and nuclear disarmament.

Albert Einstein's contributions to physics revolutionized our understanding of the universe and laid the foundation for modern physics. His theories of relativity are among the most profound and widely tested principles in science. Einstein's work continues to inspire generations of scientists and has had a profound impact on fields such as cosmology, astrophysics, and particle physics.

Albert Einstein passed away on April 18, 1955, in Princeton, New Jersey, but his legacy as one of the most outstanding scientists in history endures. His name has become synonymous with genius, and his face an iconic symbol of scientific inquiry and discovery.

The QR codes will take you to the Amazon Book order desk.

The Lost Empire of Sheba

Sam, a very famous female Archeologist has obtained information that may lead her to the location of Lost City of Sheba.

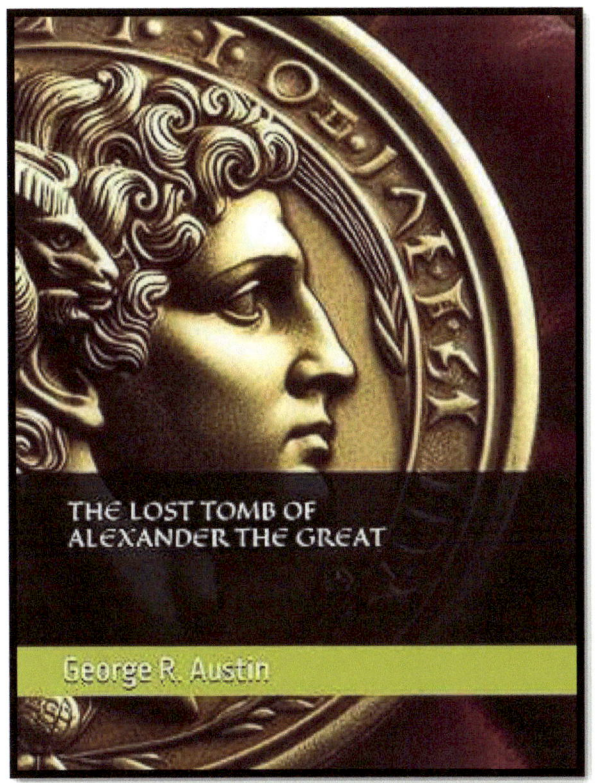

Lost Tomb of Alexander the Great." By G. R. Austin

In the heart of a gripping mystery, Sam and Flynn find themselves drawn into an international adventure surrounding the theft of priceless artifacts tied to Alexander the Great. From the bustling terminals of JFK Airport, where a discovered clue sets their journey in motion, the duo embarks on a thrilling chase across the globe. Their first lead takes them to the historic city of Cremona, Italy, to a factory where art brushes aren't just tools, but potential keys to unlocking the mystery.

As they delve deeper into the case, Sam and Flynn navigate a world filled with luxury, intrigue, and dangerous secrets. They consult with museum curators, interrogate key individuals, and encounter a cast of enigmatic characters. At the center of their search is the elusive thief known only as "The Shadow," whose cunning moves keep them on edge.

The stakes rise as clues lead them to the ancient city of Amphipolis, Greece, where a hidden treasure room reveals far more than they ever expected. A cryptic Greek riddle hints at a discovery of legendary proportions—the lost tomb of Alexander the Great. With this tantalizing clue in hand, Sam and Flynn set their sights on Egypt, determined to uncover one of history's greatest mysterie

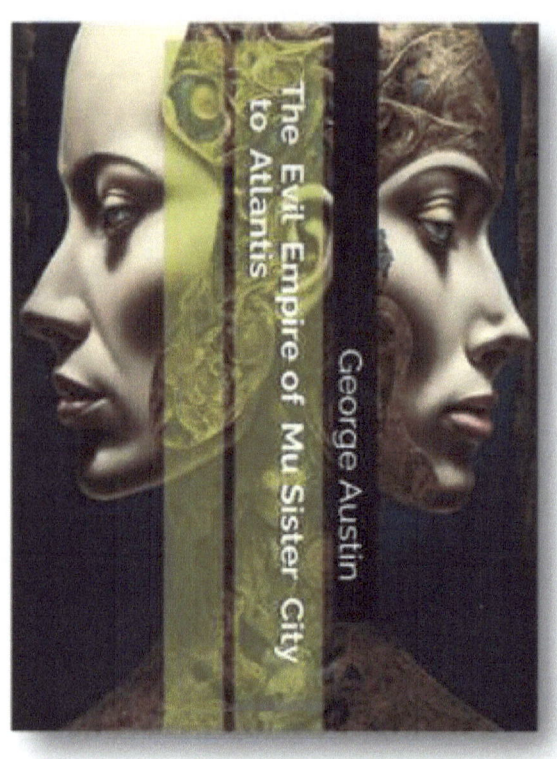

NEW BOOK
"The Evil Empire of Mu, Sister City of Atlantis."
By G.R. Austin

Join us as we embark on an epic adventure, exploring a world that was once thought to be lost forever. In this quest a Vulcanologist discovers a hidden tunnel on the Yucatan peninsula that leads to an underground city deep beneath the surface of the Pacific Ocean. It is the Hidden Empire of Mu, sister city to Atlantis.

Our motley crew consisting of a famous Archeologist, a little green alien, a 9 foot Yeti and a four foot tall Technologist probed deeper into this new venture. Together with our Vulcanologist they discover that RATONDEA DRECKBALL III, the Emperor of the Lost City of MU, has a sinister plot to destroy all human life on the surface of our planet and rule as the Emperor of the world. Can he be stopped? Only time will tell.

1980 – 1992 Directorial training/
Script writing Film Industry
Workshop Inc. CBS Studio
1964 – 1968 Acting training/
camera work Film Industry
Workshop Inc. Columbia Studio

Thinking outside of the box: Creative Days

The Legend of the Lost Nazi Gold Train

Book Four

History holds many secrets, but few are as persistent as the tale of the "Nazi Gold Train," a legend born from the chaos of World War II. Whispers speak of a train laden with unimaginable riches—looted artifacts, glittering gold, and perhaps even the legendary lost Fabergé Eggs—hidden beneath the rolling hills and dense woodlands near Wałbrzych, Poland. Despite endless searches, no irrefutable evidence has surfaced to confirm its existence. Yet, the promise of treasure and the lure of rewriting history continue to spark the imagination of treasure hunters worldwide.

In 2024, the legend finds new life. Dr. Samantha MacCambridge, a brilliant archaeologist with fiery red hair, and her dashing partner, Dr. Flynn, are drawn into the mystery of the missing train. Their pursuit takes them deep into the forests of Wałbrzych and onward to the lavish Villa Potężna Willa in Warsaw. Along the way, they uncover a labyrinth of secrets involving stolen World War II artworks and high-stakes dealings in the underworld.

With the help of Interpol's Alexander Renshaw and the enigmatic master thief Leonidas Gherkin, Sam and Flynn navigate a world where the line between history and myth blurs, determined to uncover the truth and reclaim the lost treasures of the past.

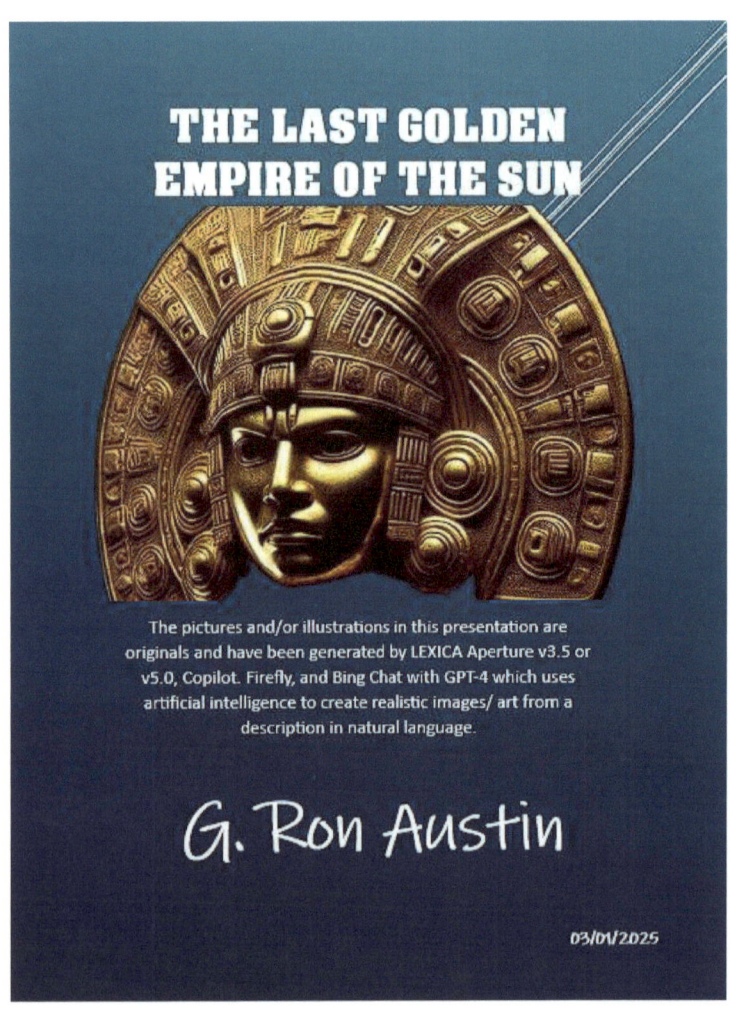

The golden bust shimmered under the museum's lights, its hollow eyes staring straight ahead, as if caught between recalling a forgotten past and longing for the treasures of a world lost to time. For a fleeting moment, it almost seemed to whisper its secrets, but silence prevailed.

Its craftsmanship echoed the artistry of the Aztecs or the Mayans. Yet, it had been found among the recovered artifacts of the elusive Nazi Gold Train, buried not in the heart of Mesoamerica but hidden within the spoils of World War II in Poland. How had it come to be there? Who had crafted it? Any hope of discovering its true origins seemed to have vanished along with the civilization that created it.

Thinking outside of the box: Creative Days

"The Lost Continent of Terra Australis" explores the enduring legend of a vast, hidden land said to exist in the southern oceans. From ancient maps and explorers' journals to modern theories of submerged civilizations, the book traces humanity's fascination with this mysterious continent, blending history, myth, and the thrill of discovery. It invites readers to journey through time and imagination, questioning what might have been lost beneath the waves and what secrets the Earth still holds."